1 Bridges, buildings and other struc

Introduction

In the past, people made their homes from whatever material was available. In time they experimented with new designs and often tried new **materials**, always trying to build bigger and better.

▼ These people lived in the woods and forests of Great Britain. Their job was to burn wood to make **charcoal**. They made their homes from trees and leaves. When they had cut down all the trees to make charcoal they would move to a new part of the forest.

▲ Small stones can be used to make large bridges. The stones are fixed to make an **arch** shape. The shape of the arch provides the strength.

▶ Building in steel is even stronger. The **triangle** shapes in this tower make it **rigid** (stop it bending). The top still moves 15 cm in a strong wind.

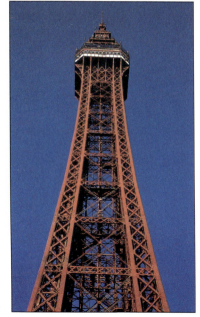

Blackpool tower

▲ Look at the triangle shapes made by the steel tubes. They make the 'Revolution' strong, rigid and safe.

Q1 How can small stones be made to bridge a large gap?	Q2 Why are steel tubes sometimes fastened together to make triangular shapes?

1 Bridges, buildings and other structures

Shapes for strength

You are going to find out how the shape of a structure can increase its strength. You may use drinking straws, LEGO parts or Meccano parts.

Apparatus

☐ LEGO Technic 2 kit 1032
or ☐ drinking straws and pins
or ☐ Meccano parts

 Take care if you are using pins!

A Make these shapes. ▼

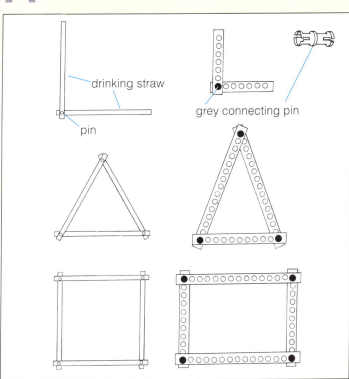

B Now gently try to change the shape of each structure. ▼

C With the square, pull and push on opposite corners to see what happens. ▼

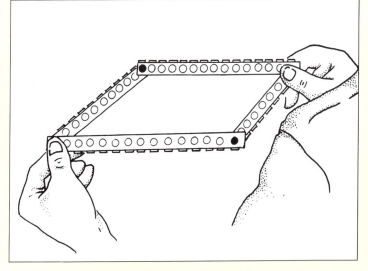

D In your book, draw the shapes before and after pushing the sides apart. ▲

Q1 Which shape was the most rigid?

Q2 In your book draw some different ways of *making the square shape rigid*. Build your best design and test it to see if it is rigid.

1 Bridges, buildings and other structures

Designing and testing bridges

Bridges have to be strong enough to carry a load over a river or railway line. In this experiment you are going to see if the design or shape of a bridge affects its strength. Each bridge must use only one piece of A4 paper and must carry the greatest weight without collapsing.

Apparatus

☐ 1 piece of A4 paper for each member of the group
☐ 2 clampstand bases or blocks of wood ☐ set of slotted masses
☐ steel washers or other test weights

Sorting out the problem
In your group decide on the width of the gap between the **bridge supports**. Use two clampstand bases or two blocks of wood for the supports.

Designing
Spend a few minutes thinking about good designs. Draw sketches if it helps. Discuss your ideas with everyone in your group.
Now choose some good designs, enough for everyone to build one bridge.

Building
Build your bridge to your chosen design. Remember to use only one piece of A4 paper for each bridge.

Testing the bridge
In your group, agree on a fair way of testing the bridges and think about:
☐ how to make the test a fair one.
☐ what to use as test weights (slotted masses, steel washers, etc.).
☐ how to load the bridge fairly until it **collapses**.
☐ how to record the results in your book.
Now you can test each bridge until it collapses.

Q1 Draw the shape of the strongest bridge.

Q2 What load did your bridge carry before it collapsed?

2 Making structures stronger

Building walls

All buildings have to be strong enough to stay up in the strongest wind. In this section you are going to find out how to make structures strong and rigid.

Apparatus

☐ LEGO Technic kit 1032

A Build two walls with LEGO bricks to make a corner. Arrange the bricks exactly as shown. ▼

B Now build another wall, but overlap (bond) the bricks. You will have to use some half sized bricks. ▼

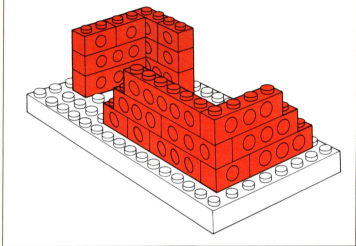

C With your finger gently rock the walls to find out which wall is the stronger. ▼

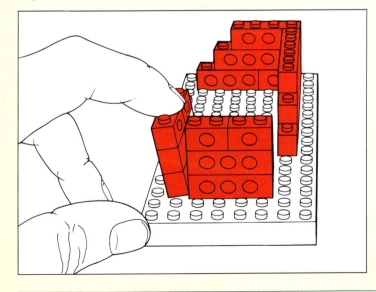

D Here are some other ways to bond bricks together. ▼

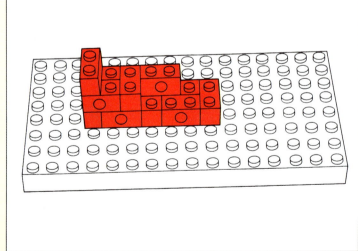

Q1 Which LEGO brick wall is the stronger?

Q2 Why do you think this wall is the stronger?

Q3 Builders use building blocks made of concrete or brick. Use a do-it-yourself book to find out other ways to bond bricks together. Draw the patterns in your book.

2 Making structures stronger

Testing shapes

A tube can be almost as strong as a solid rod and uses much less material. In this experiment you are going to test the strength of drinking straws.

Apparatus
- 2 clamps and stands
- drinking straws ☐ sticky tape
- set of slotted masses
- loop of thick string

Q1 Copy this table.

Number of straws in the bundle	Mass added to collapse bundle
1	
3	
5	

A Use two clamps and stands to set up a test rig. ▼

B Tie a loop of string at the middle of the drinking straw. Add the hanger to the string. Carefully add more weights until the straw collapses. Record the weight (mass added) in your table. ▼

C Use two small pieces of sticky tape to fasten three straws together in a bundle. You will also need bundles of five and seven straws. ▼

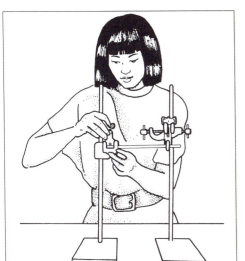

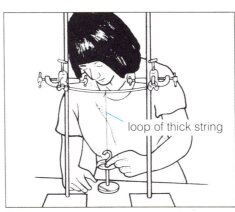
loop of thick string

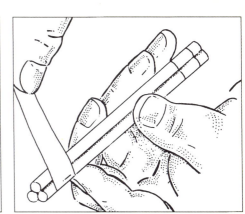

D Repeat **B** first with the bundle of three straws, then five, and finally with seven straws. ▼

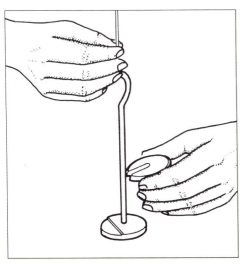

Mass (g) vs Number of straws in bundle (0, 1, 3, 5, 7, 9)

E Draw a graph of your results. ▲

Q2 Use your graph to guess what weight nine straws will take before collapsing.

Extension exercise 1 can be used now.

2 Making structures stronger

Steel and springs

Steel is an ideal building material where very strong structures are needed. It will stretch a bit under a load. We use this effect in springs. Some springs can be squashed (compression spring), others can be stretched (extension spring). Both types behave in the same way. Let's find out how springs work.

Apparatus

- [ ] 2 clamps and stands
- [ ] spring [ ] set of slotted weights and hanger [ ] pin
- [ ] metre rule [ ] plasticine

Q1 Copy this table.

Load (g)	Pointer reading (cm)	Extension (cm)
0		
50		
100		

A Fix the spring to the clamp. Put the hanger on the lower end of the spring. ▼

B Use plasticine to fix the pin on to the hanger. Put the rule in the other clamp. The '0' end should be at the top. ▼

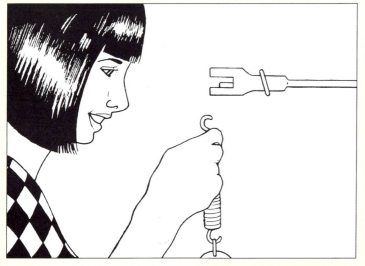
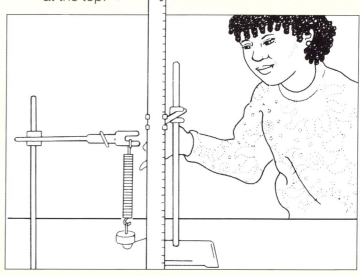

C Adjust the rule so that the pin points to a number. This is your starting point. Record the scale reading in your table. ▼

D Add another weight. Note the pointer reading. Repeat **D** four more times. Complete your table each time. ▼

E A graph of extension against load looks like this. If the graph is a straight line then we can say: 'The extension is directly proportional to the load'. This is also known as **Hooke's law**. It applies to car springs, bed springs, even bridges and the Blackpool Tower. ▶

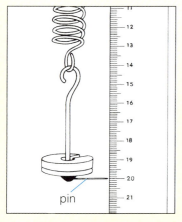
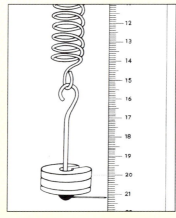
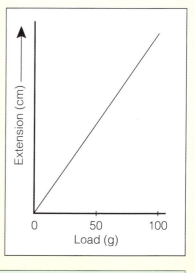

Q2 Draw a graph of extension against load for your spring.

2 Making structures stronger

Making and testing concrete

You are going to use sand, cement and gravel to make **concrete**. You will add various things to the concrete and test it for strength.

Q1 Copy this table.

Block number	Material added to concrete	Mass (weight, g) needed to snap block

Apparatus

- ☐ moulds for concrete blocks
- ☐ labels ☐ plastic tweezers
- ☐ plastic mixing bowl ☐ spatula
- ☐ cement ☐ sand ☐ gravel
- ☐ plastic scoop ☐ 2 G-clamps
- ☐ 4 lengths of steel wire
- ☐ few strands each of glass fibre and straw ☐ bucket of sand
- ☐ slotted masses and hanger
- ☐ beaker of water ☐ wooden stirring rod ☐ rubber gloves
- ☐ eye protection

Wear gloves and eye protection

A Cover the bench with newspaper. Label the moulds 1 to 4. ▼

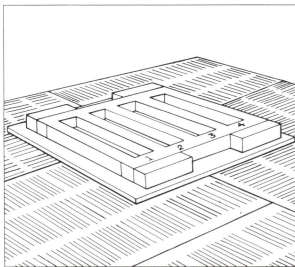

B Put on gloves and eye protection. Put four scoops of cement, eight scoops of sand and sixteen of gravel into a bowl. ▼

C Add a small amount of water and stir. Continue to add water until you have a creamy paste. ▲

D Pour a little concrete into moulds 1 to 4. ▲

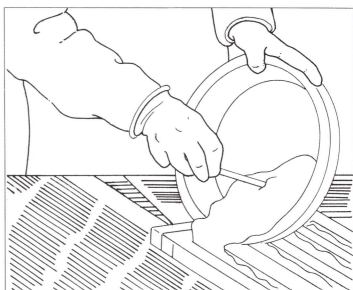

7

2 Making structures stronger

E Put a few strands of straw into mould 1, glass fibre into mould 2 and wires into mould 3. Then pour in more concrete to fill each mould. ▼

F Use tweezers to mix the strands in the concrete. Leave the concrete blocks until next lesson. ▼

G Remove the blocks from the mould by knocking gently on the underneath. ▼

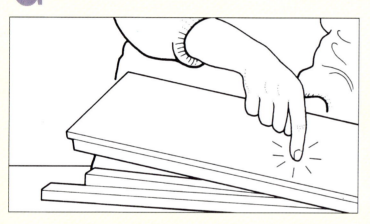

H Rest block 1 between two tables. Fix each end of the block with G-clamps. Put a bucket of sand under the block. ▼

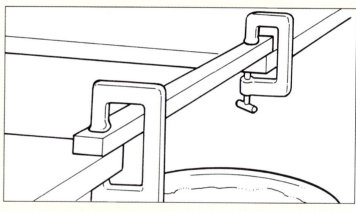

I Put the weight hanger on the block. Carefully add weights to the hanger 100 g at a time. Continue until the block cracks or breaks. ▼

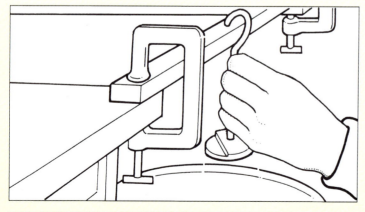

J Record the weight needed to break the block in your table. Repeat **H** and **I** using the other three blocks in turn. ▼

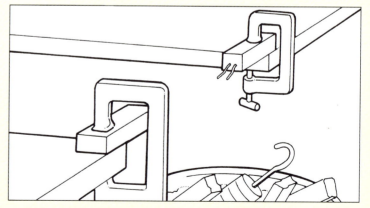

Q2 Why did you add the wire and other strands to the concrete?

Q3 How can you tell adding strands made the concrete stronger?

Q4 Adding materials to the concrete **reinforces** it (makes it stronger). Which material was best at reinforcing concrete?

2 Making structures stronger

United we stand

▶ The bones in the body of a bird must be very light in weight. They are made from many strong light tubes.

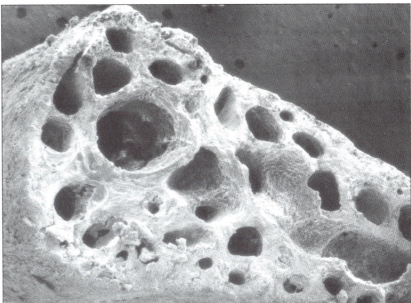

▲ The stem and branches of a plant or tree are made from many thin fibres. Each one on its own is weak, but together they are very strong.

▼ When a piece of paper or cardboard is bent into a wavy shape like this we say it is **corrugated**. The shape gives it extra strength. The savings in weight and cardboard make it much cheaper than if it were solid.

▶ Ropes are made in a similar way. A single fibre is weak but a lot of them together can be made into a very strong rope. Steel ropes carry the roadway of a suspension bridge.

▼ Civil engineers use the corrugated shape to strengthen this football stand. The beams are made from concrete reinforced with steel.

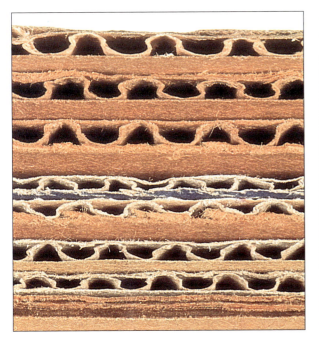

Q1 What do you think would happen to the roof of the football stand if the beams were not reinforced with steel?

3 Forces and movement

Forces can…

▶ Change the shape or size of an object.

▼ Change the speed of an object.

The turning force of the engine makes the car go faster

The pole bends

The cushion changes shape

The centripetal force of the track makes the cars go round the corkscrew

▶ Change the direction of movement of an object.

▲ Weight is a common force. It is the force of gravity. Everything is pulled towards the centre of the Earth. The bigger an object the greater the force of attraction. It is everywhere and we cannot turn it off.

air resistance up

weight force down

▶ A spring balance is used to measure force. Force is measured in units called **newtons**.

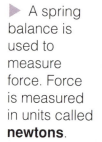
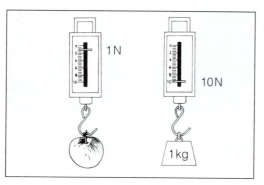

▲ Forces always act in pairs. This ice skater has two forces acting on her. The man pushes up and her weight pushes down.

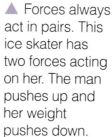

upthrust from water

weight force down

3 Forces and movement

Investigating balance

If a structure is not balanced it will topple over. In this experiment you are going to investigate forces which balance.

Apparatus
- metre rule with a hole in the centre
- 2 clamps and stands
- 3 small loops of fine string
- 2 sets of slotted masses and hangers
- write-on sticky tape

Q1 Copy this table

left-hand side			right-hand side		
mass (g)	distance of mass from centre hole (cm)	mass × distance	mass (g)	distance of mass from centre hole (cm)	mass × distance
50	10	500	50		
50	20	1000	50		
50	30	1500	50		
100	10	1000	50		
150	10	1500	50		

A Fix a piece of write-on sticky tape to a ruler. Mark the centimetre lines and number them as shown. ▼

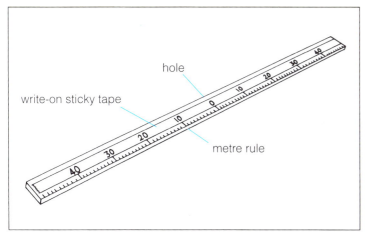

B Thread a small loop of string through the hole and hang the ruler from a clamp. ▼

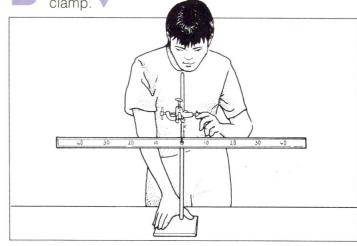

C Use the wide-open jaws of another clamp to stop the ruler from **tilting** too far. ▼

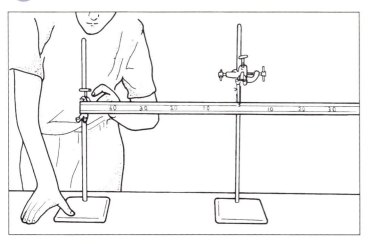

D Put a 50 g hanger on the left-hand side so that it is 10 cm from the hole. Put the other hanger on the right-hand side. Move it until the rule balances. Measure how many centimetres the hanger is from the hole, record your results in the table. ▼

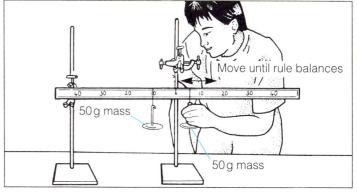

3 Forces and movement

E Repeat **D** with the left-hand mass 20 cm from the hole, then 30 cm from the hole. Record your results. ▼

F Change the left-hand mass to 100 g. Hang it 10 cm from the hole.
Move the 50 g mass on the right-hand side until the rule balances. Measure how far the 50 g mass is from the hole. Record your results in the table. ▼

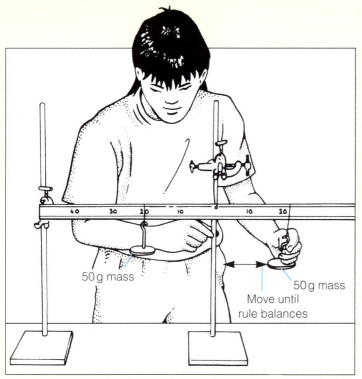

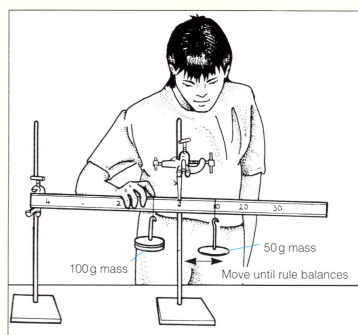

G Repeat **F** using 150 g on the left-hand side. ▶

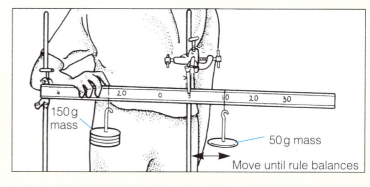

H Look at column 3 in your table. The mass and the distance have been multiplied (×) together. To complete column 6, do the same for the right-hand side. Multiply the mass (×) by the distance it is from the hole. ▼

	left-hand side			right-hand side	
mass (g)	distance of mass from centre hole (cm)	mass × distance	mass (g)	distance of mass from centre hole (cm)	mass × distance
50	10	500	50		
50	20	1000	50		
50	30	1500	50		
100	10	1000	50		
150	10	1500	50		

Q2 When a 50 g mass was 10 cm from the hole how many centimetres from the hole was the other 50 g mass when it balanced.

Q3 When a 100 g mass is 10 cm from the hole, where must a 50 g mass be placed to make it balance?

3 Forces and movement

Balancing forces

▶ A force which acts some distance from a pivot (the hole) is a **turning force**. We can find how strong a turning force is by using:

| turning force = force × distance of force from pivot |

A turning force is often called the **moment** of the force.

Examples of turning forces

Remember:
moment of a force = force (N) × perpendicular distance (m).

▼ A small force a large distance from the pivot can exert a large turning force. We use this effect in levers.

▼ A spanner is a lever. The longer the spanner the easier it is to undo a tight nut.

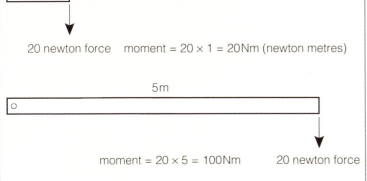

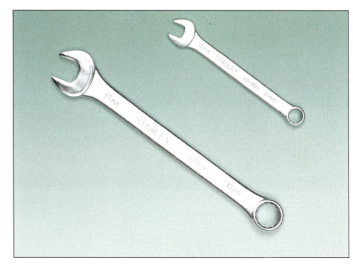

▼ Here are some machines which use levers to make our work easier in our homes, gardens, garages and tool sheds.

Q1 The cat is on the end of a long lever. It just balances the heavy weight and the mouse.

What would happen if the cat ate some cream?

Q2 What would happen if the cat turned round and walked along the lever to get the mouse?

13

3 Forces and movement

▼ If the turning moments on a structure balance it will not topple over. We say that it is a **stable** structure.

▼ This crane uses a counterweight which is placed on a short jib (arm) on the opposite side of the tower to the load. Some cranes use a different sort of **counterweight**. Large concrete blocks are added onto thick wire at the rear of the crane to stop it toppling over.

▶ Your firm makes good cranes for large building sites. Here are some facts about your crane.

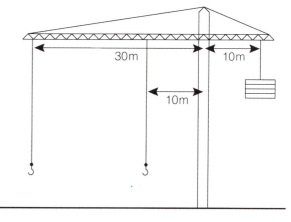

Q3 What sort of material will you make your crane from?

Q4 How will you make sure that the crane structure is rigid?

Q5 Why does the crane have a counterweight?

Q6 The crane has to lift a load of 20 tonnes when the hook is 10 m from the tower. Calculate the mass of the counterweight?

Q7 The maximum safe load when the hook is 30 m from the tower is 10 tonnes. Calculate the mass of the counterweight.

Extension exercise 2 can be used now.

3 Forces and movement

Pushing in pairs

In the following experiments you are going to find out how rockets work. The force of the escaping air in one direction makes the rocket move in the opposite direction.

Apparatus

- ☐ straw ☐ bulldog clip
- ☐ nylon fishing line ☐ 2 G-clamps
- ☐ clamps and stands
- ☐ sausage-shaped balloon
- ☐ rocket trolley kit

A Blow up a balloon and seal the end with a **bulldog clip**. Fix the balloon as shown. Tie each end of the nylon line to stands clamped to the bench. Make sure the line is **taut** and that the balloon is free to slide. ▼

B Pinch the neck of the balloon with your fingers and remove the bulldog clip. Let the balloon go. ▼

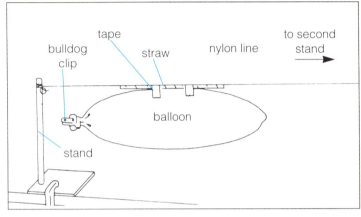

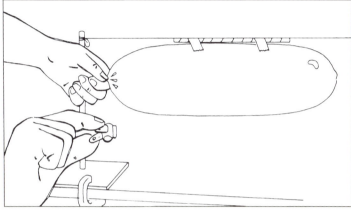

C Your teacher will set up a free running rocket trolley. It will be propelled by compressed carbon dioxide gas. ▼

D Stand back and watch carefully as your teacher hits the capsule with a nail to release the gas from it. ▼

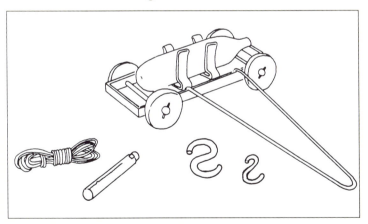

Q1 What happened when you removed the bulldog clip and let the balloon go?

Q2 Copy the drawing below. Add arrows to show which way the air goes and which way the balloon moves.

Q3 Does a force act on the balloon? Give a reason for your answer.

Q4 Explain how the trolley moves. If it helps draw a diagram.

Q5 What would happen to the rocket if the string breaks?

3 Forces and movement

Forced to move in a circle

Newton's first law says that the rocket trolley on page 15 continues to move in a straight line unless a force is applied. A force in the string pulls it round in a circle. This force acts towards the centre of the circle. It is called the **centripetal force**.

▶ If the string breaks there is no centripetal force. The rocket continues to move in a straight line. Your teacher might demonstrate this by using a match to burn the nylon line.

▼ Here are some other examples of objects moving in a circle.

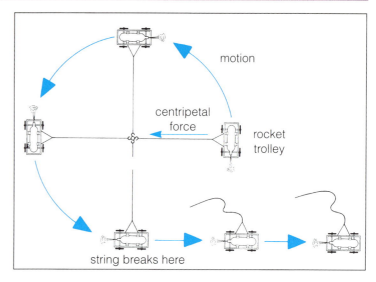

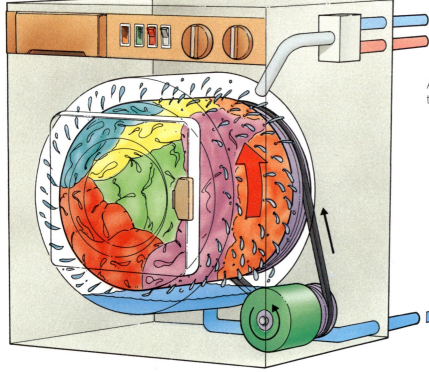

As the drum spins water is forced through the holes.

Q1 In the rocket trolley the string provided the centripetal force. Write down what provides the centripetal force in each of the pictures above.

Q2 What would happen to the motorbike and rider if the road was wet and slippery?

3 Forces and movement

Falling freely

The rate of fall of a paper helicopter

Two forces act on a falling body: the weight force which acts downwards and air resistance which acts upwards. You are going to make a paper helicopter and find out how increasing the downward force on it and reducing air resistance changes its speed.

Apparatus
- piece of thin card ☐ ruler
- scissors ☐ paperclips ☐ pen
- stop watch

Q1 Copy this table.

Number of paper clips	Time of fall (seconds)

A Draw this shape. Cut it out along the solid lines. The dotted lines are fold lines. ▼

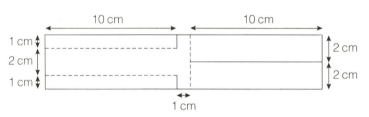

B Fold along the dotted lines to make the helicopter. Put a paperclip on the bottom. ▼

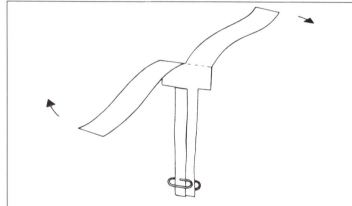

C Hold the model 2 metres above the floor. Measure how long it takes to reach the floor. Add a second paperclip and repeat **C**. Add a third paperclip and repeat **C**. Complete the table. ▼

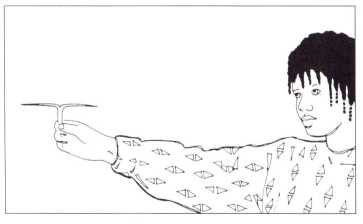

D Now shorten the wings. Watch what happens when you drop it this time. ▼

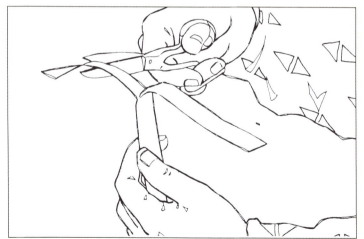

Q2 How did you increase the downward force?

Q3 When you increased the downward force did the speed increase, decrease, or stay the same?

Q4 What was the effect of making the wings smaller in **D**?

Extension exercise 3 can be used now.

3 Forces and movement

Air resistance

All objects are pulled towards the centre of the earth. As an object falls it goes faster (accelerates). Soon the air resistance limits the **acceleration**. It reaches a constant speed called the **terminal velocity**.

▶ A **skydiver** falls freely from an aeroplane. The skydiver's weight acts downwards and air resistance pushes upwards. This slows her fall. The terminal velocity is about 50 m/s (190 km/hour).

▶ At a safe height the skydiver pulls the rip cord and the parachute opens. The parachute has a large air resistance. The terminal velocity is about 8 m/s or less. The skydiver lands safely. The bigger the area of the falling body, the greater the air resistance and the slower the terminal velocity.

If there is no air there will not be any air resistance. If there is no air resistance an object will fall at an ever increasing speed. It will not reach a steady terminal velocity. This increase in speed (acceleration) is called the 'acceleration due to gravity'. On Earth the speed increases by about 10 m/s every second. More accurately measured as 9.81 m/s^2.

▶ Astronaut David Scott drops a hammer and a feather from the same height above the Moon's surface. Both hit the surface at the same time.

Q1 Explain what is meant by air resistance.

Q2 How can you make an object fall through the air more quickly?

Q3 Skydivers jump from the aeroplane at different times. They can meet up with each other during their fall. Explain how they can do this.

Q4 Why do the hammer and feather fall at the same rate on the Moon but not on Earth?

Extension exercises 4 and 5 can be used now.

4 Making things move

Energy and machines

Energy is vital to us all. Something which has energy can do work. Energy is measured in a unit call a **joule**. This is a small unit and we often use kilojoules (1 kJ = 1000 J) or megajoules (1 MJ = 1 000 000 J).

There are many forms of energy:

Heat	Chemical (including food)	Movement (kinetic)
Light	Electric (and magnetic)	Position (potential)
Sound	Nuclear	Gravitation

▼ Often we can change energy from one form to another.

Q1 Look at the picture. It contains many energy sources and many energy changes are taking place. Write down as many energy sources as you can. The first one has been done for you.

Wind – has kinetic energy.

Q2 Some things change energy from one form to another. Write down some examples and say what energy changes are taking place.

Windmill – changes kinetic energy from the wind to electrical energy.

Q3 What is the unit of energy?

4 Making things move

Ballista

In this section you are going to make things move using the energy stored in an elastic band.

Apparatus

☐ elastic band ☐ eye protection
☐ LEGO Technic 2 kit 1032 *or*
☐ Meccano kit *or*
☐ small pieces of wood or lollipop sticks ☐ plasticine (ammunition)

A In the past, armies used **siege** engines or **ballista** to destroy the walls of an enemy town. ▼

B You decide to use a model ballista as a game at the school fair. Working in your small group, design a model ballista. The elastic band will provide the energy for your ballista.

Think about:
☐ the materials you will make it from. You may choose from; LEGO parts, Meccano parts or pieces of wood
☐ how you will aim the ballista at the target area
☐ how you will change the range or distance it can throw the ammunition. ▼

C Make a model of your best design.
Decide how you will test your model. Then test your model ballista. ▼

Wear eye protection.

Q1 What form of energy is stored in the elastic band?

Q2 What form of energy is in the ammunition just after it has left the ballista?

Q3 What is the shape of the 'flight path' of the ammunition?

4 Making things move

The plastic-bottle tank

This model uses the energy stored in a **twisted** elastic band.

Apparatus

☐ plastic drinks or washing-up liquid bottle ☐ elastic bands
☐ bead ☐ stiff wire
☐ short pencil ☐ sticky tape

A Loop some elastic bands together until their unstretched length is about three-quarters the length of the bottle. ▼

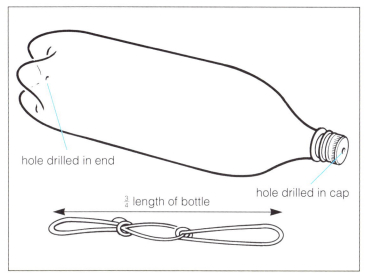

B Push the elastic band through the hole in the base. Use a small pencil and sticky tape to hold the band in place. ▼

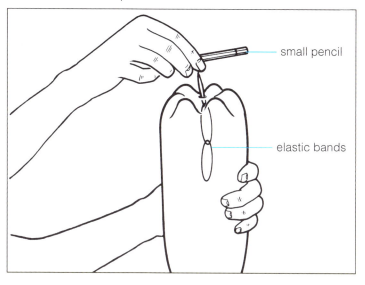

C Bend the stiff wire to make a hook. Use it to hook the free end of the elastic bands. ▼

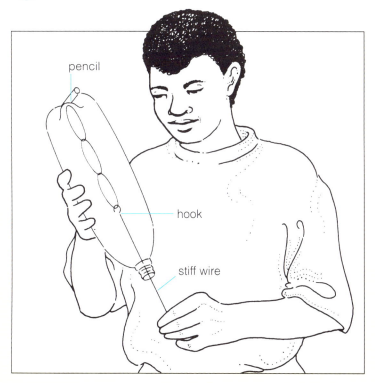

D Carefully thread on the top (cap) and bead over the wire. Bend the wire round to make a smooth drive wire.
Make sure that everything turns freely. ▼

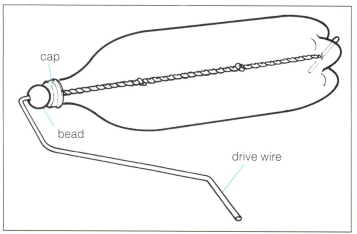

Q1 How does your tank move when you wind it up and test it?

Q2 What adjustments, if any, did you have to make to your tank or its drive wire?

4 Making things move

The plastic-bottle tank speed trials

Now that you have made adjustments to your bottle tank, let's see how fast it will go. Find a clear space about 3 metres long for your speed **trials**.

Q1 Copy this table.

Test run	Distance between start and finish line	Time taken to cover distance	Speed = $\dfrac{\text{distance covered}}{\text{time taken}}$
1st	2m		
2nd	2m		
3rd	2m		

A Mark out a start and finish line two metres apart. ▼

B Wind up your bottle tank. Put it on the start line. ▼

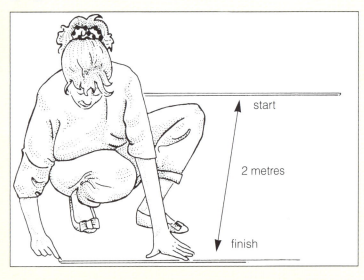

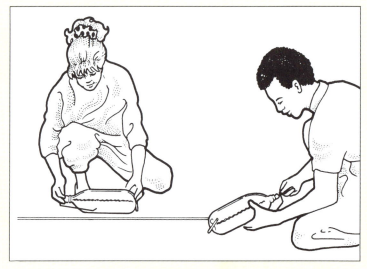

C Start the stop watch when you let go of the tank. Stop the watch when the bottle tank crosses the finish line. Record the time taken in your table. ▼

D Repeat **B** and **C** two more times.

Then use:

$$\text{speed} = \frac{\text{distance travelled}}{\text{time taken}}$$

to work out the speed for each test run. Complete the third column of your table each time.

Apparatus

☐ your plastic-bottle tank
☐ metre rule ☐ chalk to mark the start and finish lines
☐ stop watch (*or* stop clock)

Q2 How fast does your bottle tank go?

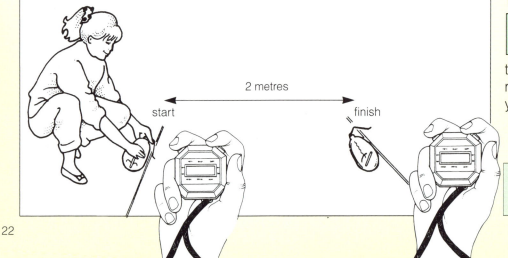

4 Making things move

Improving the grip on the road

When you were testing your tank, you may have seen it slip or skid. The tank slips because the **friction force** on a slippery surface is low. The wheel of a bicycle or car must be fitted with a rubber tyre to make sure it grips the road. The road and the tyre should have a high friction force.

▶ All racing cars have powerful engines and need special tyres to stop them skidding. They have very wide tyres made from a very sticky rubber. This gives a very good grip on the road. The tyres are smooth and do not have a tread on them.

The tyre rubber is very soft and quickly wears away. At the end of a 250 mile race the tyres are thrown away. With a harder rubber the grip would not be so good

The smooth 'dry' tyres are very dangerous on wet roads and the 'pit crew' have to change them for 'wet' tyres. These are more like ordinary tyres with a tread pattern on them.

▼ Tyres on ordinary cars must be safe on wet and dry roads. They must last for 20000 or 30000 miles. By law all road tyres must have a 1.6 mm tread on them.

▼ A mountain bike has to go over mud and wet grass. It needs a tyre with a rough tread pattern. Compare the tread pattern on the mountain bike with that on the racing bike.

Q1 Why can't you use smooth racing tyres on an ordinary car?

Q2 How could you put tyres on your bottle tank to stop it skidding? If you have time, add them to your tank and test it on the track.

Extension exercise 6 can be used now.

5 Drive systems

Making and testing drive systems

Some machines have direct drive to the road wheels. In this section we will make a model car and use gears, pulleys and chain drives to drive the road wheels.

Apparatus

- [ ] stop watch [ ] metre rule
- [ ] battery and battery holder
- [ ] these parts from LEGO Technic 2 kit 1032 [ ] 2 long wires

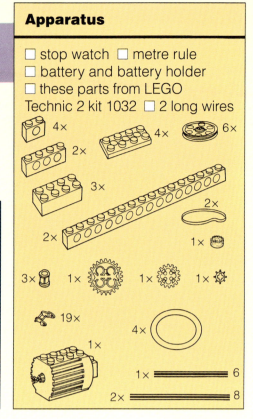

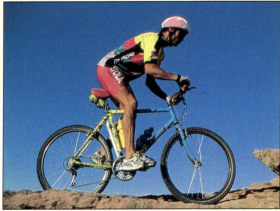

Belt and Pulley Drive

Q1 Copy this table.

Drive system	Time to travel 2 m	Easy to build
Belt and pulley drive		
Compound belt drive		
Spur gear drive		
Chain drive		

A Build the chassis of this belt driven car. ▼

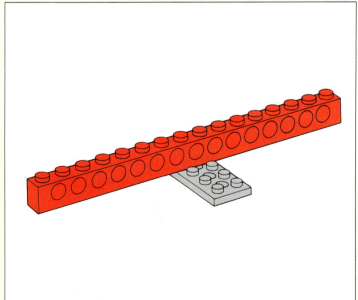

B Add the axles. They are eight studs long. ▼

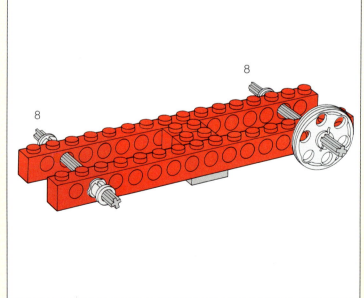

5 Drive systems

C This chassis will be used until page 26. ▼

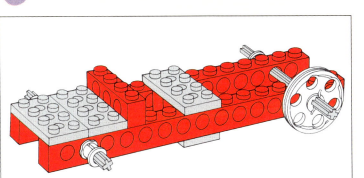

D Add the wheels, motor and drive belt. ▼

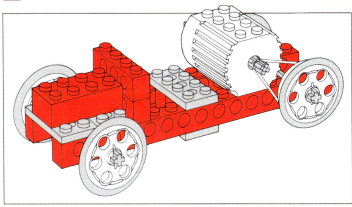

E Mark out a test track like the one on page 22. The start and finish lines should be two metres apart. ▼

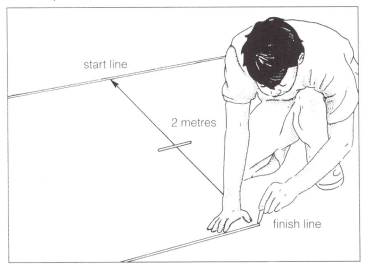

F Connect the motor to the battery unit. ▼

G Put the car about 20 cm behind the start line, and hold the stop watch. Switch on the motor. Start the watch when the front wheels cross the start line. ▼

Stop the watch when the front wheels cross the finish line. Record your results in the table.

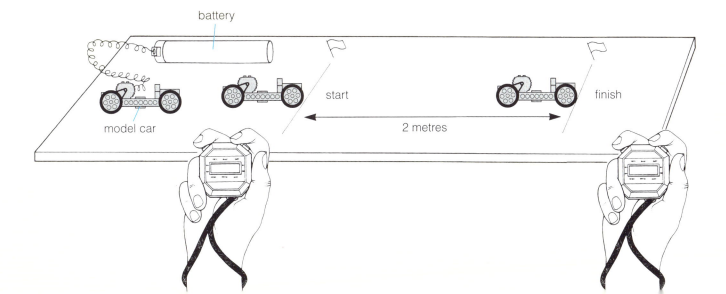

5 Drive systems

Compound belt drive

H Change to a compound belt drive as shown. Repeat **F** and **G** to test the speed of your car on the two metre test track. ▼

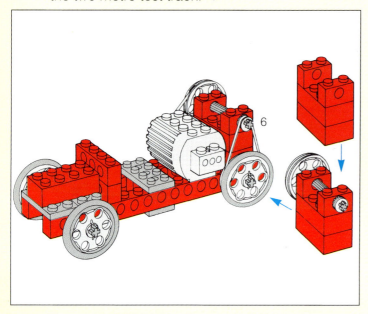

Spur gear drive

I Start with the car chassis as in **C** on page 25. Take off the pulley and put on a gear wheel. ▼

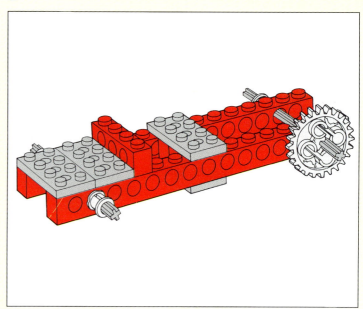

J Add the motor and small gear wheel. Repeat **F** and **G** to test the speed of your car on the two metre test track. ▼

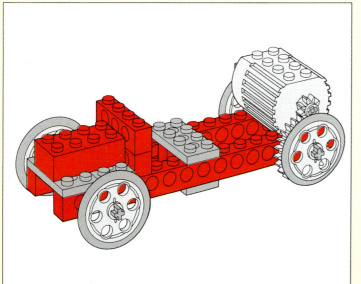

Chain drive

K Start with the car chassis as in **C**. Change to a chain drive.
Repeat **F** and **G** to test the speed of your car on the two metre test track. ▼

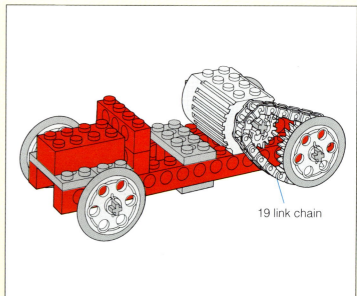

19 link chain

Q2 Which car was the easiest to build?	**Q4** Which car was the slowest?
Q3 Which car was the fastest?	**Q5** Given more time, how could you make your car go even faster?

5 Drive systems

Hydraulic systems

A liquid can transmit forces from one place to another. Car brakes use this type of **hydraulic** system. Let's find out how car brakes work.

> **Apparatus**
> ☐ 2 syringes of the same size
> ☐ syringe of a different size
> ☐ plastic tube to fit

A Connect two syringes of the same size by means of a plastic tube. ▼

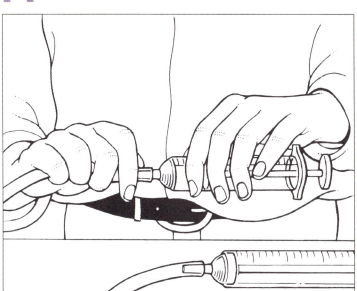

B Remove the pistons and fill the syringes and tube with water. Replace the pistons. ▼

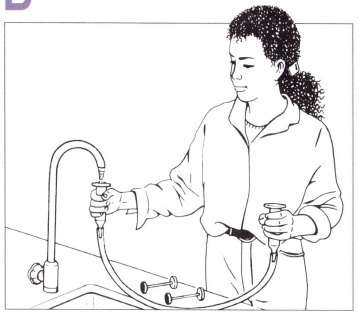

C Hold one in each hand and press the pistons with your thumbs. ▼

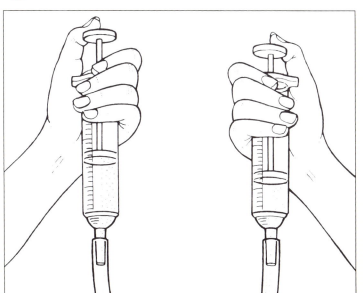

D Change one of the syringes for one of a different size and repeat **B** and **C**. ▼

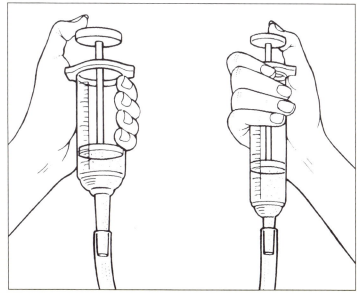

> **Q1** What difference does it make to the force if the syringes are of different sizes?

5 Drive systems

Hydraulic brake systems

Hydraulic systems like the one on page 27 are used to transmit pressure or force. The main system operated in this way is the **brake** system in a motor car.

▼ A hydraulic brake system (as shown in the diagram) has a **master** cylinder and **slave** cylinders connected by pipes. The system is filled through the master cylinder with **hydraulic fluid**.

When the **brake pedal** is pushed a piston inside the master cylinder forces the hydraulic fluid into the slave cylinders. The pistons inside each slave cylinder push brake pads onto the brake **drums** or **discs**.

If air gets into the system the brakes will not work very well. Air is **compressible** (can be squeezed into a smaller size) and therefore takes up some of the essential movement in the system.

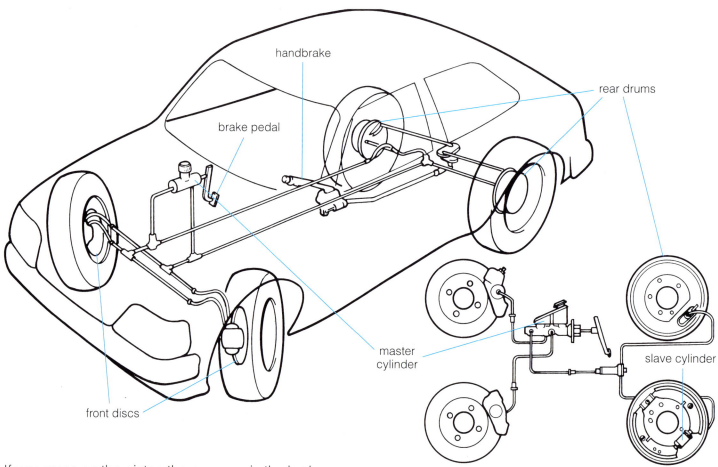

If you press on the piston the pressure in the brake fluid increases.

$$\text{Pressure} = \frac{\text{force (in newtons)}}{\text{area (in square metres)}}$$

The unit of pressure is the **pascal** (Pa).

Pressure in a liquid acts in all directions. So the pipes do not need to be straight. They can be bent to fit the shape of the car body.

Q1 In a hydraulic braking system one piston is operated by the brake pedal and the other pushes the brakes on. Which one is the smaller piston and which one is the larger piston?

Q2 Why is this?

Extension exercise 7 can be used now.

6 Using energy from natural sources

Windmills and water wheels

Long before the invention of the electric motor people used the energy in the wind and water to drive their machines. One important job was to grind cereals such as corn to make flour.

▼ To do this the sails of the windmill turned spur gears which were connected to large **millstones**. Corn was fed into a small space between the two stones. The upper stone was turned by the gears, the lower stone was fixed.

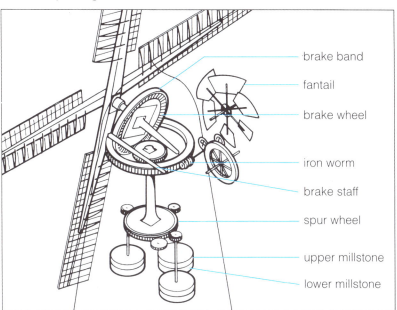

▶ The Electricity Generating Companies are now experimenting with new and very large windmills. They are very efficient and do not **pollute** the air. Usually they build a few windmills close to each other on windy hill tops. These 'wind farms' are noisy.

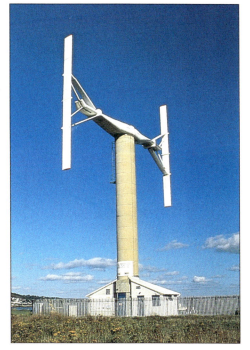

Smaller windmills can be used to pump water out of deep holes. They are also used to provide electricity for **isolated** farms. The windmill drives an **electrical generator** which charges a battery with electricity. On **calm** days the electricity comes from the battery.

Some people with caravans use a very small windmill to charge the caravan battery.

Q1 What are the disadvantages of a wind-powered machine?

Q2 In which places will it pay to use a wind-powered electricity generator?

6 Using energy from natural sources

The Savonius rotor windmill

One of the problems with ordinary windmills is that they have to point into the wind. You are going to make a **Savonius rotor** windmill. They are often used as advertising signs. They turn round if wind comes from any direction.

Apparatus

- ☐ plastic bottle (washing-up liquid or similar flat bottomed bottle
- ☐ stiff cardboard ☐ drawing pins
- ☐ 5mm dowel ☐ 2 eye hooks
- ☐ old felt-tip pen ☐ stop watch
- ☐ strip of wood ☐ hair dryer
- ☐ clamp and stand ☐ scissors

Q1 Copy this table.

Position of rotors	Number of turns per minute

A Cut the top off the plastic bottle. ▼

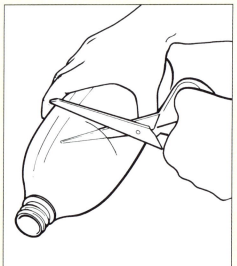

B Cut the bottom part of the bottle in half. ▼

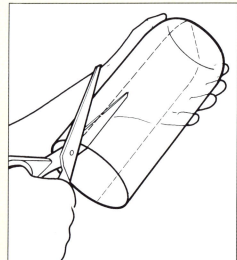

C Use compasses to draw an 8cm radius circle on the thick card. Cut out the circle. ▼

D Pin the disc to the **dowel**. ▼

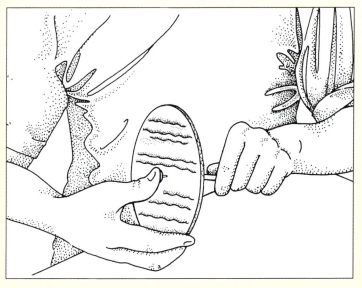

E Use the drawing pins to fix the half bottles on the cardboard. ▼

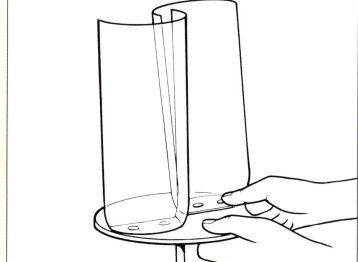

6 Using energy from natural sources

F Screw the two eye hooks into a strip of wood. Hold it upright in a clamp and stand. ▼

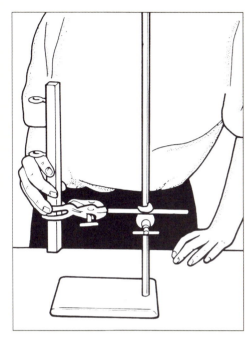

G Slide the case from an old felt-tip pen on the dowel. Put your windmill in the eye hooks. ▼

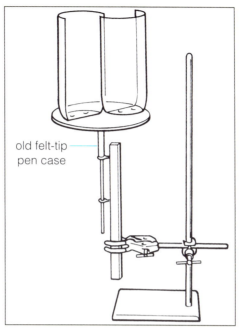

old felt-tip pen case

H Use a hairdryer or blower to test your rotor. Move the blower away until the rotors move slowly enough to count easily. ▼

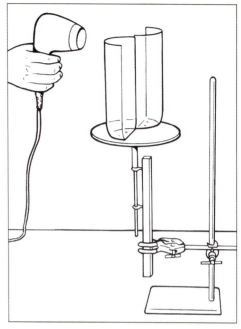

I Use a stop watch and count how many turns the rotor makes in one minute. Record your results in the table. ▼

J Move the rotors to find out what happens if you change their position. ▼

K Keep the blower the same distance from the rotor. Count how many turns the rotor makes in one minute. Record your results in the table. ▼

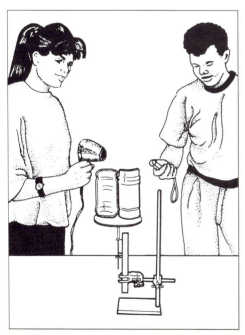

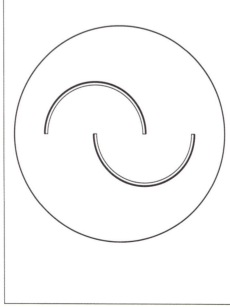

Q2 In which position did the rotor spin the fastest?

Q3 How could you make your rotor more powerful?

Q4 If you have time make and test your more powerful rotor.

6 Using energy from natural sources

Water wheels

You are going to make and test a water wheel.

Apparatus

☐ ready-drilled cork or rubber bung ☐ 2 old felt-tip pens ☐ string ☐ 8 pieces of plastic for the paddles ☐ metre rule ☐ chalk ☐ small slotted masses and hanger ☐ stop watch ☐ 2 clamps and stands ☐ these LEGO or Meccano parts: axle rod, 2 pulleys

A Push the axle through the cork. ▼

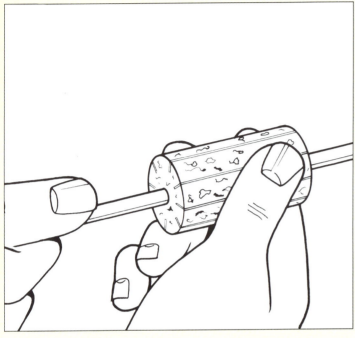

B Fix two pulleys on the axle about 1 cm apart and 1 cm from the cork. Tie a piece of string to one pulley. ▲

C Fix the plastic pieces in the slots. ▼

D Use the tubes of the felt-tip pens to hold the wheel between two clamps. Check that the wheel turns freely. ▼

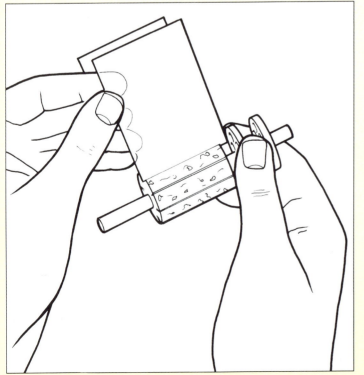

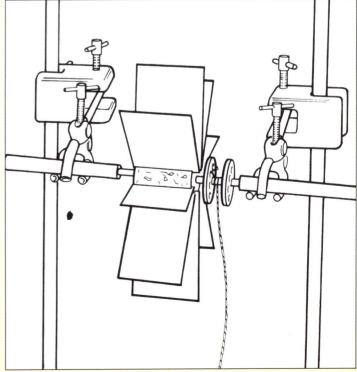

6 Using energy from natural sources

Testing your water wheel

Q1 Copy this table.

Mass of hanger = (g)	
Trial run	Time taken to 50 cm mark (sec)
1st	
2nd	
3rd	

E Put your water wheel under a tap so that the water will fall on to the paddles. ▼

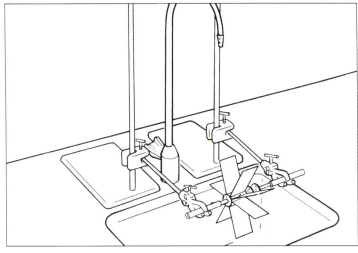

F Check that the string can move freely over the edge of the bench. Fix a small hanger on to the end of the string. ▼

G Use a metre rule and chalk to put a mark on the bench 50 cm from the floor. ▼

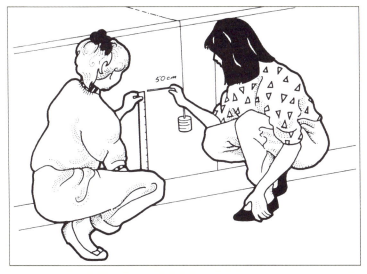

H Gently turn on the water and use your stop watch to find out how many seconds it takes to raise the hanger to the 50 cm mark. Record the results in your table.
Repeat **H** two more times. ▼

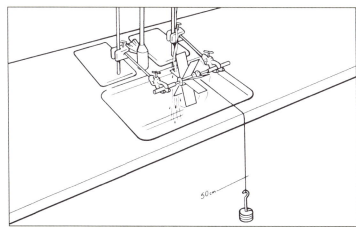

Q2 What was the fastest time to lift the hanger 50 cm?

Q3 What was the average time to lift the hanger 50 cm?

Extension exercise 8 can be used now.

7 Work, energy and power

Potential and kinetic energy

In this section we are going to find out more about energy and how it is measured.

▼ This picture shows some energy changes on a fairground ride.

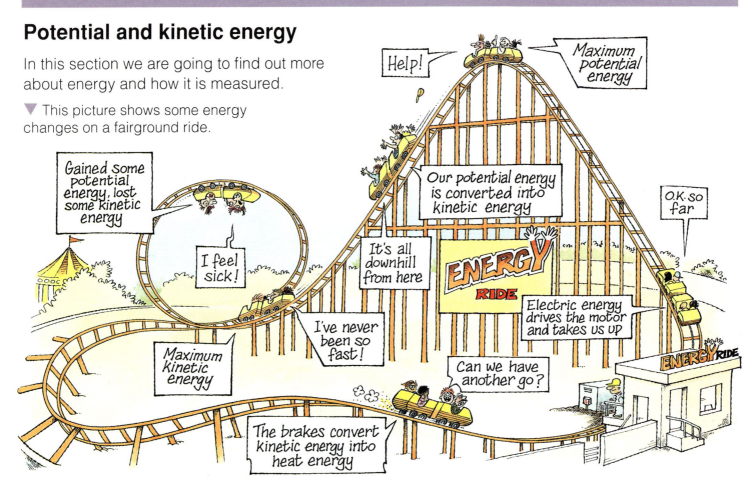

▼ An apple weighs 1 N. If you pick it up and put it on a table 1 metre high you have done 1 joule of work.

> Work done = force × distance moved (in the direction of the force)

When it is on the table it now has potential energy of 1 joule. It isn't moving so its kinetic energy is zero joules.

The apple falls off the table. Its potential energy changes to kinetic energy. Just before it hits the floor it has lost all its potential energy and has 1 joule of kinetic energy.

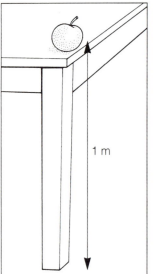
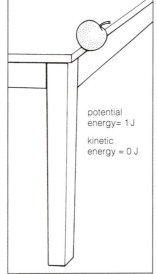
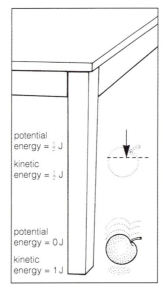

Q1 How much work would be done if two 1 N apples were lifted 2 metres?

Q2 A person weighs 500 N and stands on a diving board 2 metres above the water. What is her potential energy?

Q3 What is her kinetic energy just before she hits the water?

Q4 What do you think happens to this energy when she hits the water?

7 Work, energy and power

Kinetic energy and stopping distances

Young children often don't realise that cars need a safe distance to stop.

You need to be able to show how the speed (and energy) of a car can affect its stopping distance.

Your model car won't have brakes, but it will slow down and stop just like a real car.

Apparatus

- ☐ model car ☐ piece of wood about one metre long and wide enough for the car to roll down it without falling off
- ☐ means of increasing the height of one end of the wood
- ☐ any other apparatus you need to make your design smooth running and which will help to give an accurate result

Sorting out the problem
▼ In your group decide how to:
- ☐ give the car the same amount of energy each time you test it
- ☐ change the amount of energy you give the car
- ☐ measure the stopping distance.

Designing the investigation
Think about what measurements to take and how to record them. Discuss your ideas with everyone. Write your ideas in your book.

Carrying out your investigation
▼ Build your test rig, and use the model car to test it. When you have smoothed out all your problems, carry out your investigations. Record your results.

Presenting your findings
These will depend on your audience. Either imagine that: You are giving a road safety lesson to a group of young children and need to show them that cars can't stop 'dead'.
or
You are a driving school instructor and in a pre-driving lesson need to show learner drivers the effect of speed on stopping distances.

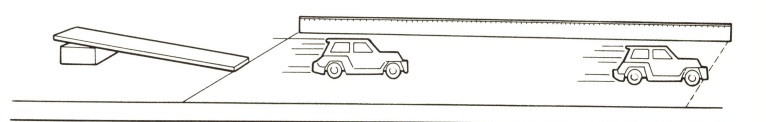

7 Work, energy and power

Kinetic energy

Anything that moves has kinetic energy. How much kinetic energy it has depends on its mass (m) and also on its speed (v).

$$\text{Kinetic energy} = \tfrac{1}{2}mv^2$$

Kinetic energy is measured in joules.

Increasing the speed has a great effect on the kinetic energy.

▼ A car travels at 10 km/hour.

▼ Its speed is doubled to 20 km/h. Its energy is now $2^2 = 2 \times 2$ or 4 times greater.

▼ If its speed is increased to 30 km/h. Its energy is now $3^2 = 3 \times 3$ or 9 times greater.

▼ At 40 km/h its energy is $4^2 = 4 \times 4$ or 16 times greater. All this energy has to be lost as heat to stop the car. Stopping distances are great at high speeds.

▼ This chart shows stopping distances for cars in good condition.

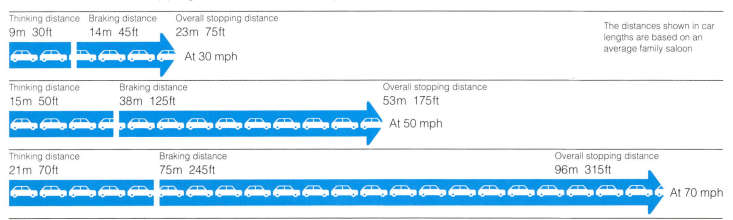

The distances shown in car lengths are based on an average family saloon

Power

This is a word which is often confused with energy or force. In science it has one meaning only. It is the rate at which energy is changed. It is measured in watts.

$$\text{Power} = \frac{\text{energy changed}}{\text{time taken}}$$

Powerful brakes change the energy quickly and stop the car in a short time.

Q1 What units are used to measure **a** energy? **b** power?

Q2 What is the difference between energy and power?

Extension exercise 9 can be used now.

8 Momentum

Rockets and momentum

In this experiment you are going to see how the momentum of the fuel from a water rocket makes it move in the opposite direction.

A Quarter fill the washing-up liquid bottle with water. Fit the plastic cap with the valve on to the bottle. Set up the apparatus as shown. If you are using a Rokit kit you will need to fit the fins onto the bottle. ▼

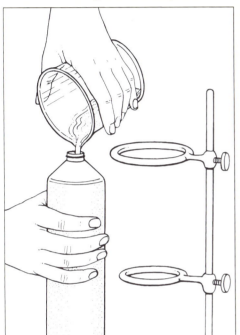
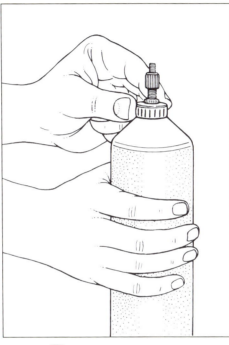

B Pump air into the rocket until it 'lifts off'. ▼

Apparatus

☐ beaker ☐ Rokit kit *or*
☐ home-made rocket made from:
☐ clamp and stand
☐ empty washing-up liquid bottle
☐ bicycle pump
☐ 2 funnel holders
☐ plastic cap fitted with bicycle tyre valve

 Your teacher will demonstrate this experiment out of doors. Keep your head out of the way of the rocket in **B**.

Q1 Copy the drawing of the bottle. Add arrows to show which way the rocket moves and which way the water moves.

Q2 What gives the rocket its upward push?

37

8 Momentum

The momentum of an object depends on its mass and its velocity.

| Momentum = mass × velocity |

When momentum is changed, a force is exerted.

▶ In your experiment the air forces the water out of the rocket. The rocket is forced forward.

Space rockets carry a supply of fuel and oxygen in giant tanks. The rocket engine burns the fuel in the oxygen. The hot gases are blasted out of the back of the rocket. The hot gases push in one direction, and the rocket is pushed in the opposite direction. Rockets are streamlined to reduce air resistance as they push through the Earth's atmosphere. In space, objects like **satellites**, which orbit the Earth at great speed, do not need to be streamlined because there is no air to push through.

Momentum in sport

There are many examples in sport where changes of momentum occur.

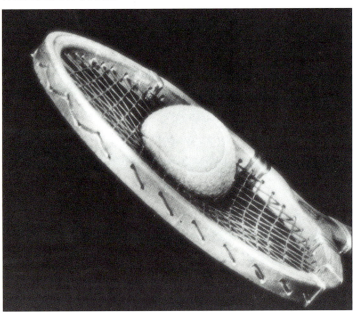

▲ The momentum of the tennis racket is transferred to the tennis ball. The ball is forced to move fast.

▶ The momentum of a cricket bat is transferred to the ball and forces it away to the boundary.

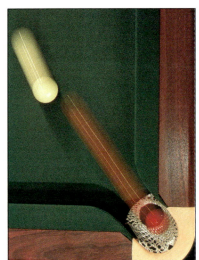

▶ In snooker, the momentum gained by one ball is equal to the momentum lost by the white ball which hit it.

Momentum is conserved in a collision. Total momentum before a collision is equal to the total momentum after the collision.

8 Momentum

Road safety: crumple zones

In this experiment you are going to see what happens to a model passenger in a head-on collision. You will then make a **crumple zone**.

Apparatus

- ☐ dynamics trolley *or* ☐ model car
- ☐ runway ☐ clamp stand
- ☐ solid support about 10 cm high
- ☐ spring ☐ plasticine ☐ scissors
- ☐ balance ☐ sticky tape ☐ ruler
- ☐ different thicknesses of paper

A Weigh out 15 g of plasticine to fix the spring to the car. Weigh out 10 g of plasticine for the 'head'. Set up your model as shown. ▼

B Put one end of the runway on a solid support about 10 cm high. Put a clampstand at the bottom of the runway as shown. ▼

C Let the trolley run about 1·5 metres down the slope and hit the clampstand base. Watch what happens to the model passenger. ▼

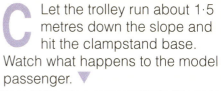

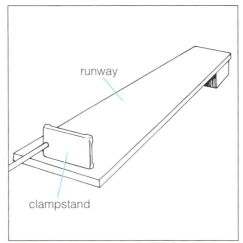

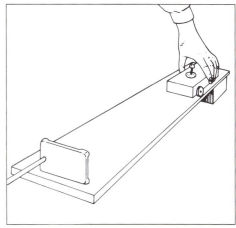

D Cut out a square of paper 10 cm × 10 cm and roll it up into a tube. Carefully fix the tube together with small pieces of tape. Take care not to crease the paper. ▼

E Fix the tube onto the front of the trolley with tape. ▼

F Let the trolley roll down the slope the same distance as in **C** and watch what happens to the model passenger.
Repeat **D**, **E**, and **F** with crumple zones made from different thicknesses of paper. ▼

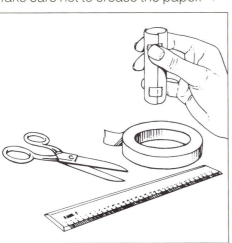

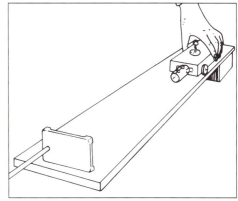

Q1 What happens to the model passenger in the head-on collision in **C**?

Q2 Does the paper cylinder make any difference to what happens to the model passenger? Why is this?

Q3 Does thick or thin paper make a better crumple zone?

8 Momentum

Momementum and road safety

▼ In a collision, momentum is changed and a force is exerted. The slower the change takes place the smaller the force. A seat belt stretches a little and increases the impact time and reduces the force. A motor bike helmet is padded for the same reason.

A car travelling at 30 m.p.h. at the moment of impact. Driver and passenger with seat belts on

A car travelling at 30 m.p.h. at moment of impact. Driver and passenger without seat belts on

▲ Cars are designed with crumple zones at the front and rear. The crumple zones increase the time of the collision and reduce the impact force. The body structure around the driver and passenger compartment is still intact.

Q1 Where is the best place for a crumple zone on an ordinary car?

Q2 Explain the effects of a collision in terms of momentum.

Q3 Explain in terms of momentum how a safety belt helps reduce the effect of the collision and reduces injury.